Dedication

To my amazing wife Tessa and our wonderful
daughters Lucy & Sophie.

Contents

Acknowledgements

Thank you to Andrew Barrow for helping me to find my energy voice.

The future of energy

1. Introduction

In this book I aim to take the reader through a possible future for energy generation, transportation and utilisation, seeking to make some bold calls on what energy will look like in 2030 and beyond. I don't seek to provide a detailed 'textbook' on the future of energy, instead I want to bring together some discussion on energy and thoughts on a range of topics which, for me, form the fulcrum of the challenges ahead of us. Although focused on the UK energy system, the ideas are transferable to any country in the world.

Energy is a huge field, touching every part of society. Without it we couldn't cook, heat our homes, make steel, travel or pretty much do anything. Since humans first made fire to warm themselves and cook, energy has been a cornerstone of progression and since the times of Watt and Brunel it is hydrocarbons in the form of coal, oil and gas which have driven us forward; forming the cornerstones of a revolution which has changed every aspect of our daily lives.

In 2020 we stand at a crossroads. On the one hand to continue our existing path reliant on hydrocarbons and the resultant impact on the planet. On the other hand, an alternative route in which we find another way, utilising hydrocarbons differently – and in lower volumes – and finding energy from 'alternative' sources including many that already exist and are rapidly moving from niche to mainstream.

There exists a huge range of information on the 'energy transition' with competing technologies and theories vying for supremacy. It's easy to fall into the trap of believing there is an easy answer or 'silver bullet' to the huge challenges we face. It is substantially more complicated, with an inevitable patchwork of future technologies rather than a single simple solution. There is no perfect answer to the challenges we face but most will, in some way, shape the way we use energy through the next decade and beyond.

There is a raft of excellent information out there on all facets of the energy transition – but what I don't see is enough open debate about the pros and cons of different options and the true potential of emerging technologies. I believe it is only through open dialogue and discussion that we will find the best path through the complex maze of challenges through which we must navigate.

Energy is a rapidly evolving and often highly politicised topic and as such this book is clearly titled the 2020 edition. Within this book my predictions will inevitably be wrong to some degree as technologies will advance at different paces and there will be ideas that may change the face of energy forever, those which haven't even been even thought of yet. I do believe it is in the debate and dialogue about the future that we will innovate and generate ideas together and create amazing outcomes.

I will be back for an update in 2021 when it will be interesting to see whether my predictions are turning out to be broadly right or whether I was way off the mark.

This book is designed to question, challenge and raise debate and as such I'd welcome your thoughts and comments at the following:

www.johnarmstrong.co.uk/futureofenergy

alternatively, you can email me your thoughts:

futureofenergy@johnarmstrong.co.uk

or contact me on LinkedIn:

www.linkedin.com/in/johnmichaelarmstrong

Scan with the LinkedIn app:

PART 1: SOURCES OF ENERGY

2. Talking carbon and greenhouses

I don't intend to dive into the science behind the climate crisis. The scientific case is compelling, and the impact will shape our planet for generations to come.

I've met several people who doubt the validity of the climate change theory and I've enjoyed several lively debates on the topic. I would however say that what matters is not individual opinion but that of society and the governments that run our countries. Opinion has moved substantially over recent years and further substantial regulatory interventions around greenhouse gas emissions are inevitable – this is hugely important to the world of energy and will have huge impacts on the way we travel, heat our buildings and run our industries.

To understand the energy transition, you need to get a grip on greenhouse gas emissions and their link to energy. Carbon dioxide is just one of several gasses which cause the greenhouse effect and is responsible for 76% of global warming. Next are methane at 16% and nitrogen oxides at 6%, both of which are emitted mostly during agriculture and industrial processes.[1]

When we talk about carbon emissions some big numbers come into play. Currently the world produces about 33 giga tonnes of carbon dioxide each year[2]. This gives a world average of four tonnes of carbon dioxide emitted per person each year. There is a large disparity across the world with for example each person in the United States being attributed 19.5 tonnes each compared to residents in the UK being responsible for a significantly lower 5.65 tonnes. With plenty of other extremes in either direction such as Qatar at 38 tonnes a person and Ethiopia at a tiny 0.1 tonne[3] there is clearly huge differences between each of our footprint on the planet.

I really like the six-tonne number as it puts into context an individual goal of a net zero[4] target by 2050 per person. This is the target for most western nations to limit the impact of global warming.

It's impossible to find an average person, or an average household, so I thought I would make it personal and share my own – here are some useful numbers from 2019 for my own carbon footprint which I calculated at carbonfootprint.com.

- Heating and electricity: 1.5 tonnes (six tonnes shared between four people)
- Flights for business (one return trip to Germany): 0.7 tonnes
- Flights for pleasure (one return trip to Belfast 0.5 tonnes
- Hotels for work: two tonnes (one to two hotel trips a week)
- Car travel: four tonnes (fifteen thousand miles in a family car)
- Rail travel: 1 tonne (typically one or two, 200-mile trips a week)
- Car replacement: around three tonnes a year
- Food: 1 tonne

- Other (clothes, insurance, computer equipment etc): 3 tonnes
- Solar panels: negative 0.25 tonnes a year (about 1 tonne for the house shared between the four occupants)
- **Total: 16.75 tonnes**

It does not take long to smash through the average of six tonnes and my halo from thinking by being a Brit I wasn't as bad as the gas-guzzling Americans is substantially dinted if not shattered.

These numbers are useful when you consider the potential for a personal carbon budget of zero tonnes by 2050. I am going to have to make some big changes or have some forced on me!

72% of greenhouse gas emissions come from energy production.[5] If you look at my personal numbers above it is clear just how much comes from transport and energy for the home. These two areas alone make up 60% of emissions for my life and I suspect are feeding into the emissions of the other areas as well, such as hotel stays. Put simply, if you don't crack energy you don't crack climate change.

Having a good conversation about carbon matters. It is easy to become 'insta-green' – all surface and no substance. But let's be clear there is nothing green about using a re-useable cup in Times Square if you just flew to New York for the photo (flight carbon

700kg[6] vs coffee cup saving 11 grams[7]). With all the over simplified messages about carbon and energy it is simply too easy to feel green when the reality is significantly different. Doing a personal carbon assessment is a great way to get a handle on your own impact. If you can't measure it, you can't control it.

Have a go at your own annual carbon review. I will come back and track mine in the 2021 edition.

If you can't measure, you can't control it.

3. Going green – Renewable sources of energy

There is no such thing as guilt-free anything – however some sources of energy are greener than others.

When we discuss renewables it's important to be clear that these are forms of energy through which no carbon is emitted during their production and utilisation. Examples of renewables are wind, solar and biomass – although biomass is a little more complicated and the subject of a later chapter.

What's the potential for wind energy?

Between 2010 and 2020 there has been a staggering increase in global wind capacity with six hundred and fifty gigawatts global installed capacity in 2020 growing by around sixty gigawatts a year.[8] To put that

wind capacity
650 GW

size into context a typical large coal power station is around two gigawatts.

In the UK, the growth of offshore wind and its cost-competitiveness against fossil fuels is credited with spurring a substantial decline in carbon emissions for electricity generation over the past decade making most of the contribution to the UK's overall carbon reduction achievements. In 2019, an impressive 20% of UK electricity was generated by wind.[9]

20% UK electricity produced by wind

The current trend in offshore wind is for turbines to get ever bigger and with more areas of the seabed being offered up for development by the Crown Estate all the time the stage is set for super growth in the sector. Globally, wind capacity is expected to go up by an incredible 112% over the next decade.[10]

Wind energy does however come with some challenges:

- It isn't always windy. There are some days where it's not that windy at all. On 5th January 2019 wind assets managed just 142MW of generation. Compare that to the windiest day in 2019, 10th December, when at peak a staggering 13 gigawatts was being produced in the UK.[11] At scale however, for example across a whole nation, wind is reasonably predictable so with the right kit it is possible to be able to plan for supply changes as the wind comes and goes, making use of other fuels as back-up or storing energy for future use.

- The wind isn't always where you'd want it. The least windy town in the UK is St. Albans to the north of London. The windiest place is the Shetland Islands. With most of the 'good' wind available offshore in the north and west that energy needs getting to the large cities… most of which are inland or in the south. Wind will inevitably form part of the drive to decarbonise electricity globally. It is however not the only answer as its intermittency and geographical challenges don't quite make it the 'silver bullet' for our energy problems.

The challenges above are offset by ongoing cost reductions for offshore wind, making the technology able to compete head-to-head with hydro-carbons.[12] With its completely green energy and an increasingly improving economic case offshore wind is set for huge growth over the next decade.

Use the sun

Solar energy has the potential to provide an abundance of energy and with 623GW of total installed capacity is the most installed renewable technology.

Demand, much of it driven in the early stages by Government subsidies, has prompted growing interest and economies of scale, making solar photovoltaics accessible to millions of homes as well as industrial scale solar farms. The total installed UK capacity is around 13 gigawatts.[13] That's the same as six or seven large coal power station at full power.

However, as green as solar is, it does come with some inherent challenges in operating the system.

- The sun doesn't shine for a lot of the day. In fact, solar very rarely runs at its peak potential in the UK which is not to say it doesn't work – photovoltaic panels run on daylight rather than direct sunshine so there's always something being produced.
- Solar panels take up a lot of space. Unlike wind which can be parked out of the way offshore, solar panels take up an awful lot of land and have specific geographical needs to run at their best. There are simply not enough south-facing roofs for all the solar panels needed. Added to that, use of land becomes a topic of strong debate as our open spaces come under increasing pressure as populations grow.

Like offshore wind, and despite the challenges outlined above, solar has huge potential as costs have plummeted such that the technology can compete head to head with hydrocarbons. A huge milestone achieved recently is the first projects emerging which do not rely on any subsidy from the government[14] - a clear sign that solar is set for fantastic growth and is a technology that is here to stay.

Are there any other sources of truly renewable energy?

Other interesting opportunities for completely renewable energy are out there such as:

Holding back the tide

The idea of generating electricity from tidal movements remains tempting. These projects have the potential to deliver around 20% of UK energy demand (government numbers).[15]

However, the environmental impact of building one remains significant. With several projects in the pipeline I believe that one must get traction and take off, the lure of constant and consistent zero carbon energy forever must surely become too much to not do something.

Going surfing

Early excitement about energy from waves has dwindled as harsh marine conditions and poor economics have seen off early contenders.
There is however some hope of developments on this technology as it has the potential to deliver consistent base-load energy supply, unlike the current major renewable contributors; wind and sun.

Drilling deep underground

Geothermal energy has some huge potential in locations with easy-to-access heat underground. The Eden Project in Cornwall, with its 5km deep bore-holes, is a fantastic demonstrator[16]. If costs of bore-holes can be reduced, then geothermal has the potential to be a game-changer for us all. The economics however don't currently work for widespread use.

Biomass & biogas

Solid biofuels already make up around 8% of energy production globally. That's a lot of wood. In many developing nations wood is the primary energy source for heating and cooking. In developed nations, wood has historically been relegated from homes but has recently seen a resurgence in power generation. The potential for biomass and biogas is covered in more detail in a later chapter.

4. Do fossil fuels have a future?

Fossil fuels present both a challenge and an opportunity in decarbonisation; they provide exceptionally good vehicles for moving and deploying energy, not least due to the substantial amount of existing transport infrastructure in place, but also their extremely high energy densities.

With increasing focus on renewable technologies, you would think the days are numbered for fossil fuels. However, that would be an easy mistake to make as even in the most ambitious projections for reducing carbon, fossil fuel-based sources of energy still account for 77% of world energy production in 2040 – it is therefore quite clear fossil fuels are with us for quite a while.[17]

Carbon capture & storage

Perhaps the most enticing technology for decarbonisation is the idea of carbon capture and storage. The idea is to capture the carbon dioxide at point of release. At power stations or industrial sites, for example and instead of sending the carbo dioxide into the atmosphere, pump it deep underground and store it under pressure. This is proven technology and could enable even the dirtiest of fossil fuels such as coal to have a future. However, the costs of the process have to-date made carbon capture and storage an enticing but unfulfilled technology.

Processes that need a lot of heat

Some things just need a lot of energy to happen. Steel and glass-making for example, along with a range of other industrial processes that require high temperature, high intensity heat. There is an argument that hydrocarbons will still need be used to support these types of processes and potentially even be rationed to ensure these processes can continue well past 2050. It is in this area that that potentially some kind of localised carbon capture and storage could be needed.

Applications where nothing else will do

Oil and gas provide the feed stocks for a huge number of processes where no alternative currently exists. Materials such as plastics need hydrocarbons as a feed stock – although alternatives are in development.

Finally, an interesting challenge of reducing natural gas production would be a shortage of helium – a by-product of the natural gas extraction process. Perhaps mildly inconvenient when it comes to party balloons but quite a major issue for Magnetic Resonance Imaging (MRI) machines in hospitals and similar applications which rely on the element to super-cool magnets.

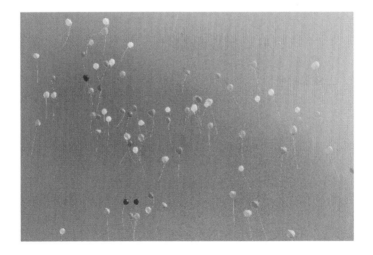

5. Fission, fusion or modular – The future for nuclear

Nuclear energy is one of the most intriguing future energy technologies, most notably because it simultaneously has the most potential to meet our low carbon energy needs, but also because it may just be the scariest and most divisive.

There are two extremely different developments in nuclear, both of which have the potential to revolutionise the way we live. One 'centralised' and the other very much 'decentralised'. They are, however, two technologies that have been "twenty years away" for the last twenty years – not least as the public's perception of nuclear has become increasingly negative over that time.

The total installed capacity of nuclear power stations globally is 460 gigawatts, about 10% of global power demand, capable of generating pretty much zero carbon electricity. There will of course be some from the mining of the core fuel, as well as the staggering amounts of concrete used to build power stations. These 450 nuclear power stations use nuclear fission (splitting the nucleus of an atom)[18] the downside of which is quantities of hazardous waste which is tricky to manage, along with processes which need to be carefully controlled to avoid exponential reactions which can cause widespread damage.

Nuclear fusion however is interesting. With fusion we are trying to replicate the reaction which happens in the sun. Put simply, if you take two hydrogen nuclei and throw them together you get helium. Unfortunately, it's just not that simple. There is however a huge amount of research into fission and it has the potential to create guilt free energy with no waste and no risk. There is one small problem and that is that no one has figured out how to do it at a commercial scale yet. It's probably twenty years away.

The next nuclear technology which is interesting is small modular nuclear reactors (SMRs). Currently most nuclear power stations have one large reactor and generate a couple of gigawatts of electricity. To house all the safety equipment and various processes around nuclear generation you need economies of scale to make it all work.

That however is not the end of the story. For decades small reactors have been used on the high seas; most notably in submarines, aircraft carriers and more recently icebreakers designed to smash through thick Arctic ice.

These small reactors present an interesting opportunity for zero carbon energy, literally on our doorsteps. Small reactors in development are typically sized around three hundred megawatts as opposed to the current larger scale reactors at sixteen hundred megawatts. These present an enticing opportunity for more decentralised power production using technology which already exists but hasn't been deployed in such a non-military or onshore application.

There are of course significant challenges away from technology development, such as the public perception of nuclear energy – never overwhelmingly optimistic at the best of times – which has waned further following incidents such as Fukushima. It's an understatement to say the idea of a nuclear reactor on your doorstep may be a difficult sell to host communities.

6. Biomass & Biogas

Solid biofuels, for example wood, already make up around 10% of energy production globally[19]. That's a lot of wood. In many developing nations wood is the primary energy source for heating and cooking. In developed nations wood has historically been relegated from homes but has recently had a resurgence in large-scale power generation.

Biomass - chopped up trees or waste wood

Biomass presents some enticing yet challenging opportunities for alternative sources of energy. I've long struggled with the idea of chopping down beautiful trees, chipping them, shipping them across the world and then burning them for electricity – although I can see the logic I probably just like trees too much. The logic of biomass generation is that the carbon dioxide which the tree absorbed as it grew would be released as the tree rots... so by releasing that carbon dioxide by burning it we are effectively skipping a step and releasing the carbon dioxide through combustion rather than rotting. The key, of course, being that you re-plant with a new tree afterwards and so the cycle continues.

In the UK biomass has been promoted heavily through the Renewable Heat Incentive (RHI) subsidy and some large biomass plants have been built.

Biogas - from rotted waste

Biogas is a very different prospect. Typically, biogas is made by rotting some form of waste such as food, sewage or cow poo, and capturing the gas – methane - resulting from that process.

Recently the largest biogas market has been supplied by capturing the gas from existing landfill sites and then burning it locally to generate electricity. This absolutely makes sense as the methane emitted from the landfill site has a far higher propensity to cause global warming than the carbon dioxide emitted once it is burned – the production of electricity at the same time makes it a double win. Biogas from household refuse or farm slurry has some fantastic potential. It does however take a lot of treatment to get it to a state where it can be injected into the main gas grid.

How is biogas made?

To make biogas the waste material is biodegraded in an anaerobic atmosphere. Done right, this emits plenty of methane. This is basically what is happening in the big green tanks you might see at the side of the road as you travel through the countryside.

Once the material is generating plenty of methane the gas needs to be treated and mixed with some other gases to enable it to be fed into the main gas grid. An alternative is to burn it locally for local heating or power generation. This saves the costs of transporting as well as treating the gas for mainstream consumption.

What's the potential for biogas and biomass?

With biogases the main limiting factor is not from the technology but from the volume of suitable waste available to process into useable gas. Future projections for the UK all show biogases having a role to play in provision of heat – but not in a significant percentage.

Biomass with an appropriate supply chain has a far higher potential. It is essential that any biomass supply chain is robust and able to deliver carbon benefit while not adversely impacting local ecosystems where the trees are grown.

The future of energy

PART 2: TRANSPORTING ENERGY

7. Storing energy - Beware hungry cannibals!

In the chapter on energy generation I touched on some of the challenges of lower carbon technologies – in that, unfortunately, many low carbon technologies generate their energy either in the 'wrong' place or at the 'wrong' time.

That's where storage comes in. Several opportunities exist to store energy for use when it's needed. These range from multi-megawatt, large-scale pumped water hydro schemes, to significantly smaller single digit kilowatt in-home batteries. All seek to address the challenge that energy is often needed at times of day when generation sources like wind, solar, tidal may not be able to meet that demand.

Storage is a broad topic with requirements ranging from 'inter-seasonal' in which energy generated in summer for example by solar, is stored for the darker winter months, to 'intra-day' where energy is stored in the daytime for use in the evening when demand is higher. Additionally, there is 'reactive' where energy is stored sometimes only for seconds to manage tiny peaks and troughs on the system and ensure the system 'frequency' is maintained. The solutions used for storage depend very much on the application required, with some technologies lending themselves to specific challenges.

Pumped water

One of the oldest forms of energy storage is pumped water storage. Put simply, water is pumped up a mountain when electricity prices are low, ready to be later released down through a generator when demand and prices are high. Examples include Dinorwig in North Wales[20] with its 16km of underground tunnels capable of generating an impressive 288MW of power when in transmission mode. Impressively the Bath County pumped storage facility in Virginia, USA, has a staggering capacity of three gigawatts and can boast that it is the largest 'battery' in the world.[21]

These types of project have huge potential, however where open 'lakes' are used they may cause substantial environmental damage if built now.

Dinorwig managed to get around the environmental impact by hollowing out huge caverns in a mountain. This kind of project comes at substantial cost and there are no new pumped hydro projects in the pipeline in the UK.

Large pumped storage projects lend themselves to longer term storage requirements across days, weeks and possible even seasons.

Large 'centralised' batteries

Recently large batteries have started to appear across the UK electricity network and there are presently around ten gigawatts of applications in the pipeline.[22] That's a staggering shift from a standing start of zero only a couple of years ago. The predominant technology is lithium ion, with containerised units set to pop up all over the country. There are several other battery technologies ranging from traditional 'lead-acid' like in a car to more expensive 'solid-state' technology used in mobile phones and potentially in the future electric vehicles.

Batteries have a huge potential to store energy local to demand in reasonable scale. Currently in the UK the economics of larger batteries are such that the sector is seeing phenomenal growth.

Small 'decentralised' batteries

In-home energy storage has potential when linked to a domestic solar panel. Some predictions suggest that across Europe installations could rise to around five hundred megawatts of installed capacity a year[23] by 2024.

Domestic batteries also have the potential to optimise income from solar panels for owners whilst limiting peak demand and supporting local electricity networks.

If there may be negative power prices on the electricity system – as we saw in April 2020[24] during the coronavirus lockdown – then batteries also give the owner, the potential to be paid to import power from the grid when prices are negative and then export when the power can be sold at a positive price.

Domestic batteries however remain costly. When I looked for my own system the payback on a six-kilowatt battery linked to my solar panels was around ten years, which felt a big gamble on whether I would stay in the same property to get payback.

One exciting area which has the potential to be quite disruptive is the use of the batteries in electric vehicles to provide decentralised storage. Potentially owners may prefer to keep vehicles 'ready to go' rather than allow power to be shared with the grid. However, they may go for it if the price is right. In the short term this technology may more likely be driven by large fleet owners – businesses and public sector organisations who have hundreds of electric vehicles parked and charging overnight.

Chemical reactions

There are several chemical reactions which can be used for energy storage. Linked to the chapters on hydrogen, one option is to use excess power to convert water in to oxygen and hydrogen. These two gases can then be stored in liquid form, converted to ammonia or compressed and stored under high pressures ready to be either combusted or put through a fuel cell when power is needed.

Any chemical reaction which can be driven by electricity i.e. through electrolysis, has the potential to store energy and these are the subject of a great amount of research.

Other ways to store energy

There are many alternative ways of storing energy under consideration such as flywheels, liquified air, compressed air, and capacitors amongst others. All have the potential to fit into the storage equation somewhere.

Beware of hungry cannibals

Batteries are an interesting technology. In a free market any form of storage technology works on the premise that prices vary, and that the storage can respond by buying when prices are low and then selling when prices are high. This is regardless of whether it's a one second 'balancing' transaction or a one hundred megawatts of energy stored in summer to sell in winter. When prices fluctuate storage can make money. The enemy of profitability for storage is flat markets where demand and supply are met equally.

Storage is an example of a cannibalising technology. The first storage to appear in the system when prices are fluctuating wildly makes the most profit. As more and more join the system each new entrant means that all participants make a little less until, ultimately, no one is making any money. In this scenario the system is however perfectly balanced.

8. A hydrogen future?

The potential for our future heating needs being met by hydrogen is increasingly widely discussed. Others suggest that increased electrification and improved efficiency mean a complex hydrogen switchover may not be needed. Sandwiched in the middle are those pushing for the middle ground, no-regrets decisions of energy efficiency and low temperature heat networks.

The one thing on which all agree is that something must be done to move from the current reliance on natural gas for heating. The recent ban on sales of new gas boilers in the UK from 2025 is a huge step[25] – however, what will replace them remains unclear.

Also, the ban doesn't address the existing housing stock which remains stubbornly reliant on natural gas, and the people in those homes unprepared or even oblivious to the fact such a fundamental change is looming large in a very short time. By comparison this makes the smart meter role out look easy!

Proponents of hydrogen believe it as the ultimate route to zero carbon heating, with trials and studies already in full swing. Others, such as the Committee for Climate Change, see hydrogen playing a potentially smaller role in the future energy system – focused only on niche sections of energy provision[26].

However, even if you believe it is technically possible, there remain some big challenges to a hydrogen switchover: whilst engineers are focused on solving the technical problems it's also necessary to explore the social, economic and environmental challenges, for example by answering the following questions:

- Who pays for a new boiler, cooker, gas fire (existing ones simply won't work with the new fuel) in every property? What if a resident wants a nicer looking cooker or fire than those currently on offer?
- How do we manage the switchover of large swathes of customers simultaneously whilst looking after the most vulnerable and limiting disruption?

- How do you deal with regional pricing? Will hydrogen users pay more or even less per unit of heat than natural gas customers or heat pump users?
- How will local air quality be managed? Hydrogen burns hotter so will produce even more harmful nitrogen oxides than gas boilers do now.

I absolutely believe engineers will be able to produce a reasonably priced domestic hydrogen boiler and that many of the safety challenges can be overcome. I think the ultimate challenges are yet to be addressed with how to enact a switchover. I suspect the factors I've identified and others I've not spotted, will make delivering a move to hydrogen at scale far harder than is currently foreseen.

I've seen many an article proudly declaring that this latest hydrogen powered boat/car/lorry is green. We need to be careful with how we talk about hydrogen as a 'green' fuel. Hydrogen is only a carrier of energy and is not in itself green. Four 'flavours' of hydrogen are used to describe how it's made and therefore how green it really is.

These are described as:

- **Black hydrogen:** made from natural gas usually in a process called steam reformation. For the chemists out there take methane, throw some steam at it and you get carbon and hydrogen. This is super carbon intensive. I read recently that to decarbonise the current global production of hydrogen used in industrial processes would need the entire renewable electricity generation of the European Union.[27]
- **Blue hydrogen:** Pretty much black hydrogen but you find a way to store the carbon dioxide deep underground through carbon capture and storage.
- **Brown hydrogen:** made through electrolysis using grid electricity to split water into hydrogen and oxygen. Electrolysis works by putting an electric current across a two metal rods submerged in the water. The oxygen travels to the positively charged rod and the hydrogen travels to the negatively charged rod. This is easy to replicate at home with a standard battery.
- **Green hydrogen:** made through electrolysis using zero carbon electricity from nuclear/wind/solar to split water into hydrogen and oxygen.

Hydrogen presents both an opportunity and a risk. Existing projects to mix hydrogen with the existing gas network present an opportunity to explore the potential of a hydrogen economy – but without a clear path to delivering green and blue hydrogen they risk exacerbating an existing problem.

9. Some like it hot – Is hydrogen the answer to those needing it a little warmer?

An often-overlooked group in our drive towards zero carbon heat are industrial processes that need higher temperatures. Examples are steelmaking (over 1000°C), glass (melting temperature over 1400°C) and even plastics recycling. Heat pumps and heat networks only get processes to around one hundred degree temperatures and certainly won't cut it for super high temperature industrial processes.

Substitution of products is an option – using less steel or more sustainable building materials for example. However, with increasing urbanisation the reality of the global economy is that there will remain a high demand for those materials used in building the cities of the future such as steel, concrete and glass.

This is where there is real potential for hydrogen as a fuel. Hydrogen burns up to a cosy 2800°Celsius (about 700°Celsius simply burned in air) giving plenty of opportunities for supporting those needing something a little hotter. It can also be compressed and stored so has potential for transport, particularly in freight.

We may be better off focusing our efforts on processes where hydrogen's potential can be realised – rather than on areas where other technologies are already proven such as domestic heating where heat pumps and heat networks already show us an achievable pathway.

The future of energy

10. Could 'dirty' ammonia be the clean fuel of the future?

The ammonia market is worth about 33 billion dollars globally, accounting for a staggering 1.8% of global carbon emissions. Produced under super high pressures from hydrogen and nitrogen, ammonia production is hugely energy intensive and more than a little bit 'dirty' when it comes to carbon emissions. Ammonia in liquid form provides potential for a green fuel as it has a relatively high energy density of 3.4kw/litre and provided it can be made from renewable electricity has the potential to be low carbon.

Could ammonia really be the fuel of the future?

I've recently been looking at hydrogen and it's clear that one of the biggest challenges of hydrogen as a fuel is its density in gaseous form along with high costs of storing and transporting it. Ammonia in liquid form provides a tantalising potential solution even if it does have a few challenges being explosive, toxic and very smelly. Potential early applications for ammonia as a fuel include shipping and maybe even aviation.

I'm starting to think that maybe ammonia has the potential to be a 'black swan' technology in some specific sectors.

A 'black swan' event is something which is highly consequential but unlikely. These are often easily explainable – but only in retrospect.[28]

How is ammonia made?

The feedstock for ammonia is simply nitrogen and hydrogen. Nitrogen can be relatively efficiently extracted from the air through cryogenic processes where air is cooled to extremely low temperatures, so the nitrogen drops separates from the other gases. Hydrogen is a bit trickier as I mentioned in previous chapters.

To create ammonia, those two gases are fed into a reactor and compressed to a high pressure in a process which has changed little in a hundred years called the Haber-Bosch process. In this process the hydrogen atoms from the feedstock are combined with nitrogen from the air. All of this is quite energy intensive as energy is needed to extract nitrogen as well as pump up the two gases to such high pressures.

What about safety and the local environment?

A quick web search on risks around ammonia gives you this:

'Exposure to high concentrations of ammonia in air causes immediate burning of the eyes, nose, throat and respiratory tract and can result in blindness, lung damage or death. Inhalation of lower concentrations can cause coughing, and nose and throat irritation'[29]

So not the nicest chemical to get exposed to. On top of the potential health impacts it's also corrosive and explosive so presents some interesting challenges to engineers.

However, engineers are well used to handling such challenging chemicals. In fact, in the United States there are already two thousand miles of ammonia pipelines along with a substantial infrastructure for ammonia manufacture and distribution. So, it's certainly not impossible that ammonia might be the right 'green' fuel for some applications.

In fact, existing hydrocarbons are distributed via pipelines, trucks, petrol stations etc. Petrol and Diesel come with their own challenges in terms of environmental and explosion hazards.

Finally, ammonia is regularly used in refrigeration and so it is highly likely you are passing near safely stored ammonia regularly in our everyday lives at sites such as hospitals and supermarkets, among many others.

This is of course all depends on being able to manufacture the hydrogen and nitrogen used as feedstocks for the process without carbon emissions. Without a pathway to zero-carbon hydrogen and nitrogen there is no zero-carbon ammonia.

Would ammonia be more expensive than hydrogen?

A study by the Royal Society[30] highlighted substantial savings using ammonia in rail, shipping and heavy road transport. In fact, nearly 80% less cost than that of hydrogen gas and a third of the costs of liquified hydrogen.

It is perhaps these three areas where ammonia has some interesting potential as a fuel for the future.

Ammonia's potential is on the side of a hydrogen economy filling in the gaps for hydrogen. For some applications hydrogen just isn't going to work and this is where ammonia may be able to step in and offer an alternative.

How green could ammonia be?

Ammonia could absolutely be zero carbon. If you can make green hydrogen either with green electricity or carbon capture and storage, and green electricity you can make green ammonia. You just need a lot of green energy. I do think the focus now could should be on decarbonising the world's existing ammonia demand. This would then provide a sound footing for ammonia to potentially be used as a green fuel.

PART 3: USING ENERGY

11. Is the future of travel all electric?

In 2019 only 3.4% of car sales in the UK were electric.[31] With further government support and 175 new models entering the market by the ed of 2020 the growth seems set to continue. The more ambitious forecasts suggest that by 2030 there could be 130 million electric vehicles on the roads globally.[32] To put that number into context there are around 1.3 billion cars on the road now[33] so even the most ambitious estimates are still only forecasting 10% in the next ten years.

There is a risk that electric vehicles are in themselves seen as green. It is important to recognise that electric vehicles are only as green as the electricity they use to charge.

Can all vehicles go electric?

Electric vehicles make sense for certain applications. Current battery technology means most electric vehicles remain suited to shorter journeys, however high-end vehicles like the Tesla Model S, with ranges in excess of 300 miles, are pushing boundaries.

Charging infrastructure is an important barrier in take-up. The electric vehicle experience is inherently less flexible than petrol without substantial network investment use will remain limited to second cars or those using vehicles for reliable journeys.

Some of the larger 'fleets' are looking to electric vehicles, companies such as Amazon which have ordered one hundred thousand vehicles.[34] It will be interesting to see how they cope with reduced range on traditional vehicles and changes to ways of working.

What about freight?

Road freight is unlikely to go fully electric as current battery technology is not capable of moving large payloads over long distances. Even with significant improvements in technology, freight is unlikely to go electric for a long time.

There are some interesting trials of overhead line technology like that used by electric trains[35] where instead of having a battery a 'pantograph' is raised above the lorry to electric wires over-head. These provide an interesting if currently largely unproven route to the decarbonisation of freight.

As with many energy applications the answer will likely lie in hybrid options. With heavy goods using a mix of alternative fuels and batteries depending on the application.

What about other forms of transport?

Rail has a clear pathway to decarbonisation through electrification. Existing electrification projects have however been dogged by overspends, with scope reduced.[36]

Could electric vehicles become obsolete?

There is the potential for electric vehicle infrastructure and electric vehicles to become obsolete. In the event that an alternative fuel, such as hydrogen, could be found then the benefits of a new tehcnology could effectively leapfrog electrification. In this case it is possible that electric assets could become 'stranded' as users move to more convenient options.

How are we going to charge all these electric cars?

The challenge we are yet to really experience is that of charging millions of electric vehicles. Electric vehicle chargers use a singificant amount of electricity and electric vheicles are only green if the electricity they are using is green itself.

In order for electric vheicle growth to continue, the ability to charge vehicles in the right places is the bigger challenge. The automotive industry has already priovided the vehicles – it's now for the infrastructure to follow.

The future of energy

12. The future of flight – Three energy futures

The future of air travel just doesn't seem clear – right now there is no readily available and technically proven alternative option for air travel.

Every minute 84 flights take off somewhere in the world, with more than four billion journeys being made by plane each year[37]. Those flights consume nearly three hundred million tonnes of jet fuel annually, making up around 2% of total global CO_2 emissions[38]. Before the advent of Covid-19 and lockdown protocols worldwide, air travel was expected to double over the next two decades.[39] We will have to see if – or, probably more likely, how quickly – those numbers return to pre-outbreak levels.

Recently I've been excited to see small electric planes taking off such as the Eviation Alice electric aircraft currently under development by Eviation Aircraft of Israel.[40] Using battery technology taken from the automotive sector, these planes have been able to travel reasonable distances on one charge – and carrying a couple of a passengers. There have also been some bold statements about having electrically powered large aircraft by the end of the decade. These innovation steps however are nowhere near to decarbonising a long or even a short haul flight.

The fundamental challenge of decarbonising flight is the energy density of the storage. The physics of batteries seem to work for smaller applications but not necessarily for bigger aircraft. Currently lithium ion batteries can store around 250Wh per Kg which is thirty times less dense than jet fuel.[41] So, the weight of the batteries ends up limiting the ability of larger planes to even get themselves off the ground – never mind carry a payload.

Research has suggested that for battery powered aircraft to work the energy density would need to be nearer to 800Wh/Kg so nearly triple that of best available technology today. At the current rate of technology improvement, this kind of energy density isn't going to be available until well after 2050.

Even if the improved energy density could be achieved a huge challenge remains – that of charging large airliners on the ground in a timeframe acceptable to carriers and airports alike. To put it into context, a Boeing 747 needs around sixty megawatts of power for cruising.[42] A Boeing 747 currently has a turnaround time of 150 minutes and a 737 for the likes of Ryanair twenty-five minutes.[43] The infrastructure required to charge such a large battery so quickly would be quite incredible never mind charging a thousand flights a day at Heathrow.[44]

Even if the energy density challenge can be resolved there is still a long way to go to be able to charge the aircraft once it's on the ground.

Finally, biofuels have some potential even if they are unlikely to prove game-changing. Currently biofuels supply about 0.1% of global aviation fuel and despite some early enthusiasm haven't seen high levels to growth.[45]

Looking to the next decade what does the above mean for air travel?

The development of smaller emissions electric aeroplanes may mean we see an explosion of smaller regional airports with pilot-less air taxis. The shorter available distances will mean that short hop aviation may become a real thing. Also, smaller planes will be able to fly at lower altitudes avoiding congestion.

With weight becoming critical in range calculations, removing the pilot starts to make sense – accelerating autonomous flying to enhance the economic case.

Electricity demand at transport hubs will become an increasing challenge. With an increase in electric vehicles the demand on local electricity infrastructure will increase – Heathrow, for example has 51,000 car parking spaces.[46] Just charging that many vehicles will take a huge reinforcement of electricity provision. Assuming slow chargers (three kilowatts) and about one tenth of vehicles on charge at any one time, you'd need a fifteen-megawatt connection just for the vehicles – never mind the additional load if you wanted to start charging giant airliners in quick turnaround times. With smaller electric planes there will be no need to use giant hub infrastructure like now – operators will be able to move to cheaper smaller airports where charging provision is more easily provided from local solar and storage, for example.

For larger aircraft, alternative fuels such as hydrogen with its higher energy densities may make more sense than electrification as the energy density of these fuels will make such a switch over more logical. The development of regional hydrogen centres presents opportunities globally however, as right now no one is commercially flying on hydrogen. There is still a long way to go.

In conclusion, current available technology means the only way to reduce carbon emissions is to fly less. There just isn't a clear enough pathway to lower- or zero-carbon flight. Smaller electric planes may supplement existing routes currently served by smaller aircraft such as the Skybus to the Isles of Scilly in the United Kingdom. Without significant regulatory intervention and some big technological leaps, they won't replace our existing hydrocarbon hungry fleet any time soon.

13. The four 'C's of the energy transition

When looking at the energy transition I believe there are four lenses through which to make energy decisions. We are moving from linear decision making in an age of energy abundance into a multi-angled challenge which constantly evolves over time.

Carbon intensity – what's our carbon footprint?

Carbon intensity is a complex challenge which is not only hard to nail down in the moment, it is even more difficult when you project into the future.

Energy problems need inputs and outputs. In the case of electricity, the carbon intensity of those inputs can change each half hour, never mind five years in the future. Since 1990 the carbon intensity of the electricity grid has fallen by 38%[47] and continues to reduce in all future energy scenarios I've seen. I do think a lot of these scenarios tend to understate the impact of a huge shift to electrification and the capability of renewable sources to respond to rapidly escalating demand.

That said, gas networks have their own plan for reducing the carbon intensity of the gas in the pipes through biomethane and hydrogen mixing, so the future may not be all electric.

Capacity – can we get to the energy we need?

Even if we can get an abundance of low carbon energy (i.e. hydrogen or electricity) the next challenge comes in getting it to the point of use.

There are not insubstantial challenges with local and national grid capacity in a number of energy scenarios. These directly impact both individuals and companies seeking to make energy decisions.

That means at a residential level not everyone on a street can have a twenty kilowatt fast electric car charger. If delivery companies quickly electrify their fleets the grid connections needed to provide even overnight charging will be phenomenal.

Heating UK homes at peak time requires about 360GW[48] of instantaneous energy. For comparison current 'peak' electricity demand is about 50GW. Any shift to electrification of heating will massively increase demand on both national and local grids. Even hydrogen may not present the perfect solution – 'green' hydrogen uses electricity to generate it and huge volumes of storage would be required to enable enough to be there on super cold days.

Cost – how much will it cost us?

Carbon and capacity both come at a cost. Resilient, zero-carbon options don't always come cheaply. With so many alternative options, delivering lowest instantaneous and lifetime cost presents a complex and diverse challenge.

Geography is also likely to become far more important. With hydrogen deploying locally and grid re-enforcement relying on substantial infrastructure investment, it may simply be that the costs of energy are influenced equally by where you live or where a company is situated – even down to a parish level in the UK. A little like the two million off-gas grid properties in the UK who currently have no choice but to opt for more costly liquid petroleum gas, we could see differences depending on hydrogen availability or even local electricity grid capacity.

Cooperation – how can we make this happen?

Finally, and most importantly the answer to the above three challenges is in cooperation. We simply must look to our neighbours to make better use of what we do have. There are plenty of fantastic examples out there; we are just going to have to work harder and think differently to get there for example by:

- Considering whole system planning and local energy systems across zones – enabling capacity to be managed at a more local level.
- Using neighbours' waste heat such as data centres, sewage or air conditioning to optimise performance of heat pumps.
- Understanding local demands on networks – such as electric vehicle charging prior to requesting giant grid connections – i.e. do neighbours charge their vans overnight to enable delivery, meaning their capacity could be used during the day for office workers?
- Shared storage through heat networks enabling reduced demands on the system and energy sharing.

Cooperation increases the number of 'no-regrets' decisions. Energy saving and energy sharing typically always make sense … and finding ways to limit demand and peak capacity requirements reduces cost for everyone.

This type of thinking takes an unprecedented level of coordination at the local, regional and national level. Success comes in a constant open dialogue about future and current needs – and maybe giving up a little control for the greater good. The ultimate technology split is still a long way from being decided but what is clear is that **carbon, capacity, cost and co-operation** are the cornerstones of any decision-making around energy systems.

The future of energy

14. Fifth generation heat

Heat networks are an interesting proposition for decarbonisation. Government forecasts suggest around 18% of heat demand could be provided through some form of heat network by 2050[49]. Heat network technology is progressing rapidly, with the emergence of a new fifth generation approach that certainly requires consideration.

What do we really mean by fifth generation?

The fundamental premise of fifth generation heat networks is that distribution temperatures drop substantially to around twenty-five degrees Celsius. Typically, fourth generation systems run at around fifty to sixty degrees.

At this temperature the system integrates with heat pumps, enabling energy sharing – for example using the 'low grade' heat energy emitted from cooling systems which can then be redistributed round the network.

To achieve high efficiencies, data helps to constantly optimise the total system – modifying the network temperatures and flows to provide the optimised position for all users. Whereas existing networks may have summer and winter temperature settings, the fifth generation constantly adapts to deliver optimum temperatures, driving the coefficient of performance upwards - how efficiently heat pumps convert electricity into cooling or heating energy across the system.

Fifth generation heat networks have some unexpected and positive consequences over other heat decarbonisation options which matter when it comes to building future smart cities – most notably in freeing up roof space to live in with no more big cooling rejection units from air source heat pumps and ensuring local cooling effects don't make our city centres more than a little chilly. They also avoid local emissions issues which a switchover to hydrogen still won't achieve.

I didn't even know we had been through third and fourth generation. What did I miss?

The trend in heat networks has been to reduce system temperature. Older networks use steam to transmit heat, with more recent systems using water at around eighty degrees. Thinking from academia and industry has recently seen a shift towards fully integrated heat systems with temperatures as low as forty degrees in networks - with issues such as Legionella control resolved more locally to the consumer.

In practice fourth and fifth generation networks are likely to merge. Unlike your phone you don't get a little logo in the top left corner of your screen saying you are using fifth generation heat. Fifth generation is something that can happen quietly in the background with the end user not really knowing.

What is happening to make fifth generation heating technology mainstream?

There remain some significant challenges with fifth generation technology. How you deliver a system for the 'greater good' and share the carbon and cost benefits will present some interesting challenges.

There are several case studies in operation now which are showing some interesting performance. One is the research being led by London South Bank University amongst others, around the Bunhill[50] cluster in North London which features some great examples of using 'low grade' heat – the underground, sewage and cable tunnels are already being looked into and the potential really is out there.

There is also some really interesting research from the likes of Lot-Net where universities are working together to deliver some fascinating insight into how fifth generation heat networks can and will work.

What will need to happen to make it work?

Collaboration is key to fifth generation networks: parties need to look to share the carbon savings and produce a system where everyone benefits. Opportunities for energy sharing exist in many places and it is only through open dialogue that these can be turned into reality.

Also, systems need to be designed 'future proof'. Lower temperature networks and heat pumps need systems that rely on much lower temperatures. A good example is using underfloor heating instead of radiators (or higher capacity radiators). Some simple choices like taps and show heads with a low temperature differential can make a huge difference down the line.

PART 4: THE FUTURE

15. How can cities take control of the energy transition?

The energy transition presents cities with unprecedented challenges in planning for the future. Electric vehicles will place an exponential demand on electricity infrastructure, gas is being phased out and we need increasingly stringent controls on local emissions, particularly in urban centres. Simultaneously, governments are introducing an array of regulations and reporting requirements along with an equivalent, and similarly confusing, selection of subsidies.

Where are we today?

Cities of all sizes need to draw a line in the sand and assess where they are today: assess their current consumption, what infrastructure assets already exist, how is energy already used.

Taking this to the next level then gets interesting... what is the capacity of the existing systems, how many parking spaces are there that are sustainable for future electric vehicles, how do traffic flows look now, how long are people spending in the central business district? Importantly capacity maps for all the main utilities are essential – identifying bottlenecks now can reap huge benefits later.

This is where data comes in. Amazing datasets exist showing city energy consumption, people movements, traffic flows, satellite heat maps, underground infrastructure. Bringing these together can be bewildering but will help construct a 'digital twin' over time – enabling exponentially better planning.

Consider the city's direction of travel.

Cities around the world are declaring climate emergencies as well as seeking to address local pollution. This is a great time to take stock and establish a clear direction of travel.

Firstly, electric vehicles need to form part of any plan – soon a robust charge point network will be essential in all urban centres. A fast charge point uses twenty-two kilowatts so a handful of charge points in one location can quickly push local energy infrastructure to its limits. There is also a temptation to ignore residents without dedicated parking spaces. Currently a lack of available infrastructure for those without their own drive or dedicated parking space effectively locks many citizens out of joining the electric vehicle movement.

Cities typically have huge asset bases themselves, with hundreds of energy hungry buildings. A clear strategy is needed for building fabric, energy provision and consumption which takes a leadership position. Realistically natural gas is going to be around for a while, however the city may want to decarbonise quicker, it may have ageing assets that are due a replacement, or they may need a little more resilience. In which case some decisions may need to be made sooner rather than later.

Now is the time to be thinking about upping electricity distribution capacity ready for rapid expansion both for electric vehicle charging and heat pumps across urban centres. This change is happening as we speak, and demand will likely rapidly outstrip supply.

Look to your neighbours.

The energy transition is all about collaboration. Using the available datasets mentioned above, opportunities to share energy can easily be identified. Take a look around you for collaboration opportunities – do your neighbours have excess photovoltaic panels or spare heat? Such as data centres which eject huge amounts of heat in the process of cooling computer servers. Alternatively do neighbouring businesses and government buildings need to charge electrical vehicles overnight for example with delivery vans whereas commuters may need to charge theirs in the day?

The future energy world is all about collaboration – network pinch points, particularly electrical, are going to drive us together into local problem solving. City representatives can play a huge part in facilitating and even leading that collaboration.

Use energy scenarios.

Consider the cities strategy; for example, what are your plans for the next ten, twenty-five, fifty years? Even ten years feels an awfully long way away so scenarios can help. Cities are in a unique position that they can play the long game... unlike businesses with shareholders to satisfy and a need to chase return on investment, cities work to different objectives.

Energy trends around the electrification of heat, electric vehicles and digitalisation are clear – there is a lot of talk about hydrogen but realistically this is beyond the planning horizon for most of us. That said, now may be the time for cities to make systems at least 'hydrogen ready' in some way.

Take small steps now in the direction you want to go.

The thousand-mile journey really does start with a single step. There are some 'no regrets' decisions; solar photovoltaics will generate carbon-free electricity for twenty-five years; energy efficiency always makes long-term sense and should always be the 'first fuel' in this conversation. Electric vehicles are only going to grow in popularity, with a resulting increase in localised energy demand. Local interventions to reduce nitrogen oxide and particulate emissions are simply the right thing to do from an environmental, health and safety and even a child protection perspective.

For complex interdependent infrastructure the array of challenges can seem bewildering. Some cities have made some clear steps – enforcing ultra-low emission zones, supporting heat network infrastructure, replacing diesel busses with electric alternatives. I've been particularly impressed with some of the bold moves made to address infrastructure challenges in cities like London, Copenhagen and Bath.

Finally, cities need to look for support and need not wait for a clear plan. As I said at the beginning of this chapter there are a sometimes-bewildering array of subsidies out there. Globally cities are doing some incredible things; the C40 cities are a fantastic example.

16. Energy efficiency – The two billion tonne challenge

We simply cannot wait for 'silver bullet' technologies to take away our problems when we have the technology now to begin to do something about it. Through building an efficient, low temperature, digitally linked system, significant decarbonisation of heat is achievable.

To accelerate the decarbonisation of heat we must:
- Ensure new buildings are highly efficient and future-ready, accelerating work to improve existing housing stock.
- Focus on digitalising the energy system to make it truly 'smart'.

- Grow 'no regrets' options such as efficiency improvements, heat networks and waste heat recovery.

Where are we now?

About half of energy consumed is used for heating, half of which is used for heating buildings and providing hot water; contributing 22% of UK greenhouse emissions. That's one hundred million tonnes of CO_2 per year.[51]

There are about thirty million homes in the UK, of which 83% rely on natural gas for heating[52]. It's interesting to note that in 1970 it was only 43%. Gas heating has grown in a world of abundance and low costs, leaving us with a legacy of poorly insulated homes contributing more than five tonnes of carbon per home per year to our collective emissions. The technology exists today to easily halve these emissions which would equate to more than two billion tonnes of CO_2 unnecessarily emitted between now and 2050 if we just carry on doing what we are doing now. And that's just using the domestic figures, never mind commercial.

To date, significant steps have been made in reducing the carbon intensity of power. In fact, three quarters of the UK's reduction in carbon emissions since 2012 have come from the power sector.[53] In April/May 2020 the UK spending a month without generating any electricity from coal recently demonstrates just how far we've come.

These improvements mask little improvement in heat and transport. With electrification of transport progressing, albeit too slowly, heat remains the final frontier of the epic energy transition which we are currently experiencing.

We now need to urgently address the challenge of decarbonising heat which will require a transformational change to the energy system, not to mention a cultural and societal shift if we are to get the buy-in of tens of millions of people. Through investments in energy efficiency, digitalisation and networks now we can make a change like that achieved in the electricity generation industry in the preceding decade. Through making strong choices now we can and must achieve a step change in decarbonisation like that achieved in power.

Energy efficiency

So where do we start? Energy efficiency matters. Before we discuss how to generate 'heat' in different ways it is essential to look at how we can use less and how we can future proof systems to be able to work well with future technologies.

House efficiency is measured using an alphabetical rating from A to G. A typical 'A' rated house has energy costs of around four hundred pounds a year compared to over two thousand for a 'F' or 'G' rated dwelling. Most properties in the UK are rated 'D' with consumption around double of an 'A' rated building[54].

Some recent analysis by the smart thermostat manufacturer 'Tado' showed UK houses performing on average the worst in Western Europe with an average drop of 2.5°C at a zero ambient temperature over five hours – compared to around one to one point five in other countries.[55]

By 2050 95% of homes will need to be 'A' rated – that means improving to the energy efficiency of 15.5m homes between now and then at a rate of around 0.5m a year.

The real key to this is that heating properties inefficiently typically uses higher temperatures. The recent study by the consultant Ramboll on converting the energy system of the town of Cowdenbeath to different systems highlighted that in all scenarios investing in energy efficiency and running at lower temperatures had benefits in all scenarios.[56]

New funding mechanisms are required now to incentivise rapid investment in energy efficiency.

New build houses

New build has an important role to play. Each year a further one hundred and forty-three thousand new houses and flats are built which could easily rise to three hundred thousand.[57] New buildings should be built with the highest levels of efficiency. Banning domestic gas boilers from 2025 is a strong step in the right direction and will encourage investment in alternatives – with further investment bringing down the cost of technologies such as heat pumps. Like in offshore wind where industry and government backing saw costs reduced substantially and rapidly as the industry grew.

Establishing high efficiency, low temperature systems in these properties will pay dividends later.

Low carbon heat networks

After energy efficiency comes a further 'no regrets' option, particularly where there is a high density of heating or cooling demand. Significant savings in energy consumption can be achieved where heating and cooling are situated in proximity. These seek to create a 'breathing' network where heat is used again and again as it passes through the city environment. Academic research such as the Lot-Net program is making excellent progress in the area of understanding how these 'breathing' networks can be established.

One cornerstone of these networks operating well is the integration with the end consumers. Where good building insulation and low temperature systems smooth peak demands and lower overall temperatures. These enable systems to tap into 'waste' heat such as sewage, rivers and tunnels enabling heat pumps to achieve highly energy efficient co-efficient of performance values. A high co-efficient of performance means lots of heat is produced for very little electrical input.

Heat networks also provide a significant storage opportunity. In times of high demand networks can store significant volumes of energy as well as potential using 'phase change' storage. This can make a huge inroad into the challenge of smoothing the peak energy demand.

What matters is that networks become smart – providing grid balancing through utilising their immense storage potential.

Heat pumps & hydrogen

Heat pumps have the potential to provide a replacement technology to gas boilers. They can play an important role in transitioning to a low carbon system. Hybrid heat pumps, which use gas to meet the occasional peak in heat demand offer a potential 'no regrets' option. These are a good option for retrofit but it is worth highlighting that installations need to have two key features. Firstly, the heating system needs to be improved to operate at lower temperatures and secondly, the system needs to be 'smart' to limit peak demand.

Evolution of the new-build market is extremely relevant to lowering costs and preparing for mass roll out of heat pumps into the retrofit market. The Government's Future Homes Standard will play a key role in supporting the deployment of heat pumps at scale in new-build properties, providing confidence to the supply chain to invest and innovate to meet this demand.

Hydrogen remains a high cost option, requiring either large volumes of very low carbon electricity or carbon capture and storage. Of the options that can be deployed at scale today, it won't be viable for some years to come. Should we wait for hydrogen at a time when action is required today? In ten years', time, when the technology might be ready, we will have already emitted well over half a billion tonnes of carbon from our housing stock unnecessarily.

What will the system look like to support the transition?

Heating is distinctly seasonal, with peak demand being five times higher than peak electricity demand - needing around three hundred gigawatts of energy to be delivered. In comparison, the electricity grid can deliver around sixty gigawatts at peak. Whether you use heat pumps or hydrogen this is a challenge to over-come. Technologies such as batteries, storage and smoother consumption can make a huge difference.

Clearly the electrification of heat will increase demand significantly over time, but a significant amount can be done to meet this expectation. Through building an efficient, low temperature, digitally linked system, electrification of heat is achievable.

A call to arms

I believe the challenge of the heat transition can be achieved if we:

- Invest in future-ready and highly efficient buildings which operate with lower demands and temperatures whilst accelerating improvement of our existing housing stock.

- Support heat networks and other low temperature systems – those that can utilise sources of waste heat.
- Invest in making the system 'smart' through smarter grids, smart meters and systems to facilitate time of day pricing, load shifting and storage.

To do this we need to act now to incentivise and encourage rapid growth in energy efficiency, digitalisation and low carbon heat sources. With the right environment we can absolutely achieve rapid decarbonisation of heat over the next decade and replicate the successes already achieved in other sectors. We have the technology now to take big chunks out of the two billion tonne challenge – so why wait? The earlier you start the less you emit. A goal of zero by 2050 is great – but we mustn't miss the clear fact that savings now matter. We can do so much more by acting now... so why wait?

17. Will energy still come in vanilla?

Gas and electricity are a pretty 'vanilla' product. Carbon intensity of gas and electricity is typically reported as an annual average and unless you're a large commercial consumer, kilowatt hours are priced at a flat rate regardless of time of day or even year. Of course, neither of these scenarios are reflective of the true nature of the energy flowing through gas pipes and electricity cables, situations in which the carbon intensities and costs are constantly changing based on millions of complex inputs and outputs to the system.

Carbon intensity - the ice cream

Looking at carbon intensity of the grid we can see just how much this varies during the day[58], even your location in the country has an impact, not to mention the variables now stretching to include having my own solar panels or other forms of onsite generation.

Where a supplier buys its gas also matters... Statoil's Sleipner gas field off the coast of Norway limits CO_2 from production through capturing carbon in the rocks below the sea bed, whereas liquid natural gas cooled (using a lot of energy) and shipped from a long way away will have a much higher carbon intensity. So, where the gas is bought from really does matter.

The ability to achieve the best carbon outcome in any given period is impossible to calculate in the moment. There are simply too many variables. However, in the future technology should help us dynamically choose how green we want to be in any given moment.

Price - the cone

Having decided on how green I want my carbon I can then look at price. If I have photovoltaics on-site then this is likely to be my best option. So, I need to prioritise when it is sunny. But what if I have a battery and I'm expecting prices to go up later in the day? As mentioned above; only large consumers have what's called 'time of day' pricing. With technology there is a view this will come to all sectors to manage peak demand.

For gas there are peaks in the network where gas storage is used at a cost to suppliers, to prop up demand which is equivalent to about three hundred gigawatts of energy at peak. Again, currently there isn't a lot of time of day pricing for gas, however it is likely to increase.

So, do I want to use my energy at the most cost-effective times, or do I not really mind and just want convenience? Could I cap my costs so at least my system avoids winter pricing extremes?

Customer service - Chocolate/strawberry sauce or a flake?

So how do I want to be served? Do I want to talk to someone directly or am I happy to just be served online talking to a chatbot (is artificial intelligence so good I don't even know the difference?). And then there are the extra bits. What else comes with your energy – such as smart controls, boiler maintenance or other exciting additions?

What does this mean for the future of energy?

In the future artificial intelligence and smart controls can start to make a lot of the above decisions for us. Like the way you can select the type of route you want in a sat-nav, a smart system links together consumption, local generation and supply data to provide the optimum outcome based on consumer wishes.

These kinds of considerations apply to both commercial and domestic environments. Economically, investment in energy performance tends to stack up quicker for commercial users. However, domestic users can be quicker on the uptake for the latest trends as decisions are taken in a very different way. It's going to be exciting to see how we stack up our energy in the future.

18. Thinking about resilience

At 4:52pm on Friday 9th August 2019 a lightning strike hit a transmission circuit 4.5km from the Wymondley substation in Hertfordshire. It was a normal day in the UK transmission system, but that bolt of lightning caused significant disruption. Trains were stopped for hours and the country took a significant time to recover.

'Prior to 4:52pm on Friday 09 August Great Britain's electricity system was operating as normal. There was some heavy rain and lightning, it was windy and warm – it was not unusual weather for this time of year. Overall, demand for the day was forecast to be like what was experienced on the previous Friday. Around 30% of the generation was from wind, 30% from gas and 20% from Nuclear and 10% from inter-connectors'
Report into the events on Friday 09th August.[59]

The lighting that hit that power line turned off the power to more than one million people. Although power was restored within an hour the impact went on late into the evening and was particularly felt in the rail system where one fleet of trains couldn't reset themselves when the power returned. This caused significant disruption and discomfort for those stuck on the trains waiting to be rescued.

Was it wind and solar's fault?

Much was made in the press of the changing nature of the electricity grid and the impact of more generation embedded in the system.

Logically more decentralisation should build in greater resilience as large single points of failure disappear. However, it is clear from the August power cuts that these key single points still exist. In such a complex interdependent system small disruption can still have a much broader impact. Although in this case the lightning strike caused a very short-term impact – it was the lack of resilience of certain critical elements to voltage drops which had the largest effect. Had the trains been able to be restarted by the drivers, for example, the impact would have been significantly reduced.

In thinking about energy resilience it's important to not use just past events to plan. In many ways the root cause is of less importance than the impact. A quick read of the list of global power outages shows the broad range of causes which can impact the system. Many of which are beyond our control:

- **Weather events** – If you look at large blackouts around the world many are caused by large storms, lightning or ice build-up on power lines.
- **Solar storms** – According to scientists, space weather impacting computer systems and electricity grids is inevitable.[60]
- **Technical failure** – Sometimes equipment does fail, like in the London blackouts in 2003[61].

- **Third party intervention** – A digger hitting a cable or a cyber-attack.
- **Pandemic** – A pandemic prevents energy workers from attending work to maintain the system.

So, what can we do?

All the above events and many more can cause electrical systems to shut down. You can't plan and prepare for every eventuality, however what you can do is test systems for recovery and assess potential impacts. Following the August power cuts, it's important for organisations and individuals to consider:

- What is the impact of even a short interruption to key supplies (power, water, gas etc) to our activities and our customers?
- Is there any way to physically test how we would recover in the event of a loss of critical supply?

- Can we look for ways to build in natural resilience into our systems? Renewable technology gives us chance to build in resilience, for example through local energy systems.

As the system evolves and further decentralises, considering our own more local systems becomes increasingly important. A decentralised system can be both more and less resilient in different ways and users need to consider in such a rapidly changing environment how best to mitigate the impact of system disruptions.

19. How will UK energy look in 2030?

Predictions are fun, aren't they? Deep down you know your boldest shot will be wrong – but that doesn't stop us all from trying. With energy decisions lasting decades, attempting to predict the future is an essential art.

Ten years ago Matt Cardle sang the Christmas number one and David Cameron had just started his six years in office. The UK government predicted in its base case energy scenario that in 2020[62] electricity generation would be delivered with 75 TWh from coal and 52TWh from renewable sources. The reality today is more like one terawatt hour from coal and 127TWh from renewables. And given the recent zero coal run, coal generation is likely to be even lower.[63] This epic underestimate demonstrates just how wildly wrong we can be over a decade. The pace of the demise of coal in the UK has been massively under forecast year on year as has the expectations of growth from renewable sources, mainly offshore wind. Looking back to our predictions a decade ago is important in forming our views on the next ten years as it shows just how quickly things can change when the mix of regulation, technology and market forces are right.

If anything, I believe this gives fire to making bolder predictions in terms of decarbonisation in the future rather relying in any way on the continuation of the status quo or only technologies deployed today. It does, however, make it a little difficult to pin down a view of the future.

The one consistent in any future view is a drive to continued carbon reduction and in all outcomes below, the drive to reduce carbon is assumed rather than debated.

So here we go. I suspect and hope this may open a significant amount of debate and challenge!

The decarbonisation of heat will gain speed with substantial consumer take up of air and ground source heat pumps.

Heat currently accounts for half of energy use and consequently a high share of emissions[64]. Heat pump take-up remains sluggish, however the recent ban on gas boilers is only the start of a number of regulatory interventions to drive changes in consumer behaviour around heat.

I would predict that by 2030 at least one third of homes in the UK will be heated by heat pumps. This will be driven by regulatory intervention as well as changing consumer perception of methane gas as an environmentally damaging option over electricity.

In city centres something different will be needed as there won't be enough space for ground loops, and because air source heat pumps will take up much needed roof space and cause localised cooling.

Enhanced insulation and improved efficiency will gain a sense of urgency and pace.

Currently 14 million properties in the UK have an energy efficiency rating of 'D' or lower[65]. That basically means they use twice as much energy as an 'A' rated property. It is important to note that UK properties are some of the worst performing in Europe.[66]

These properties need better insulation (roof, wall and floor), new windows and lower temperature heating systems. To meet climate targets half a million of these properties need a deep improvement every year between now and 2050. After a slow start getting our heads round this, I'd expect at least five million of the fourteen million 'D' rated properties to have been fixed by 2030.

There will remain a stubborn core of properties on outdated technology – even with recent leaps forward I was amazed to see that in the EU there are half a billion radiators without a simple temperature control valve.[67] That's just incredible when you think of the energy saving potential of such a simple change.

Hydrogen will develop in pockets as well as for transport for example for trains, shipping and heavy goods vehicles.

Regional projects like Hy-deploy in the UK will see pockets of intense hydrogen deployment. Whilst in these areas piped hydrogen to properties will be a reality, the slow rollout will mean lots of properties electrify with heat pumps before widespread hydrogen becomes a reality. Hydrogen will more likely be used for road haulage and trains where its high-density energy storage enables it to be the leading low carbon option. Where and how the hydrogen is made has the potential to become a political issue – with higher electricity costs, using green electricity to make hydrogen will become challenging – pushing producers towards 'blue' or 'brown' hydrogen made from cracking methane with steam reformation a not-so-carbon-friendly process unless capture and storage is available.

'Zombie' gas grids increase – pushing costs on those unable to switch away

As the electrification and efficiency improvements gather pace the number of people disconnecting from the gas grids will increase. If you take the above numbers for heat pumps and efficiency, then by 2030 there are seven million less gas users and those who remain on are using substantially less. The network costs remain to be paid, pushing up costs on those who can't disconnect for economic or technical reasons. By 2030 I'd be expecting us to be making a call as to whether to switch off the gas networks in certain high cost regions.

Tidal barrages – one or two will happen. . . but the capital costs and local impacts will make further projects difficult

The idea of generating electricity from capturing the tide remains tempting. These projects have the potential to deliver one fifth of energy demand.[68] However, the environmental impact of building one remains significant. With several projects in the pipeline I believe that one must get traction and take off. The lure of zero carbon energy forever must surely become too much to not do something?

Batteries – expansion will continue but will be stalled by resource

Batteries are the ultimate cannibalising technology. The more batteries on the system the less profit any of them can make. Batteries also have a limiting factor in terms of the minerals used to make them. Already electric car manufacturers are struggling to source enough lithium and cobalt and these constraints will push up costs.

The current trend is away from decentralised batteries towards larger units at key nodes in the system. I'd expect this to continue as local demands increase to support the network.

The system will get smarter and more local

Quietly the system is getting smarter. If you'd have said ten years ago, we'd be asking Alexa to pop the heating on in the office we'd have said that was something out of Star Trek. But here we are... connected devices are everywhere and quietly our energy systems are getting smarter. These little bits of inter-connectivity slowly work to improve the system one small step at a time. It may feel slow but as each device interconnects the ability of the system to make smart choices increases. Consumer demand for smarter systems has the potential to enable a smart energy system as a by-product to other much more fun functionality such as asking Alexa jokes over breakfast.

I'd expect by 2030 domestic energy to be smart in some form in most properties. Heat pumps will come pre-connected to the internet and whatever replaces gas boilers to be smart in some form. To meet demand the electricity network will need to get much better at limiting peak demands – through linked devices much of this is achievable.

That said, resilience remains a concern, with system failures becoming more common due to increasingly poor weather. Local energy systems become more prevalent as a way for communities and organisations to take control of their own energy destiny, moving away from central grids to locally managed electricity and heat networks, possibly even disconnecting from the centralised systems completely.

Heat networks will become sharing networks

Heat networks are predicted to supply 18% of the UK's heat by 2050.[69] My prediction is that they could do more. High density areas such as city centres have the potential to share energy much more than they do now, by taking 'low-grade' heat from sewers, trains, tunnels, data centres and cooling systems to use in heating. Fifth generation heat networks will use a mixture of technology and lower temperature systems to enable users to feed in and take out energy from the system. Once these systems are established the benefits of connecting will be significant and will result in a rapid take-up of the technology as networks expand.

By 2030 I believe all city centres will have some form of sharing 'fifth generation' network or be working towards installing one.

Transport – half of vehicles will be electric by 2030

I'm going to go big here. My prediction is that by 2030 half of vehicles will be electric, with all non-electric cars banned from city centres. Recent campaigns by the likes of The Times newspaper[70] have highlighted the impact on health from particulates and nitrogen oxides. Recently the number of cities tightening controls on vehicles has hugely increased with Bath, Bristol and the City of London taking substantial steps to reduce polluting vehicles in city centres.

I think this will continue and the take-up of electric vehicles will be driven not by concerns over carbon emissions but more from local air quality driving polluting vehicles away from our urban areas.

I also believe that with autonomous vehicles the rise of the likes of uber and an increasingly 'rental' or shared economy, car ownership will drop significantly. Once you can hail a cab reliably for the same cost why do you need to own your own vehicle and have the hassle of parking it?

Air travel will be less common

Air travel remains a significant challenge. I just don't see a technology yet that can get the weight and size of an aeroplane into the sky without using a significant volume of hydrocarbons. I *can* see a world where frequent flyer is no longer a badge of honour and instead frequent flyers are increasingly penalised. I'd expect air travel to at least halve by 2030, with increasing regulatory intervention to penalise those who pollute the most.

How about a black swan?

Finally, all the above is written from a current view of the world. What's the black swan event that derails this completely? Was offshore wind back in 2010 a black swan that was hiding in plain sight? Nuclear fusion, deep geothermal or exponential development of hydrogen all have the potential to significantly change the playing field. This would of course change everything I've said above.

Time will tell

Only time will tell if in 2030 the UK will have:

- Five million properties moved from 'D' to 'A' energy performance through deep retrofit.
- Ten million domestic ground and air source heat pumps installed.
- Two major city centres with a low temperature fifth generation heat network.
- Hydrogen in the pipes for one or two regions with up to two million houses connected.
- Half of vehicles on the roads electric and a halving in car ownership.
- One or two large tidal lagoons.
- A declining number of gas grids or an increasing zombie affect hitting costs for the remaining few connected.
- Half as much flying as we do now - even without the impact of Covid-19.

So, what do you think? How far off am I? Have I been bold enough?

I will come back and track these in the 2021 edition.

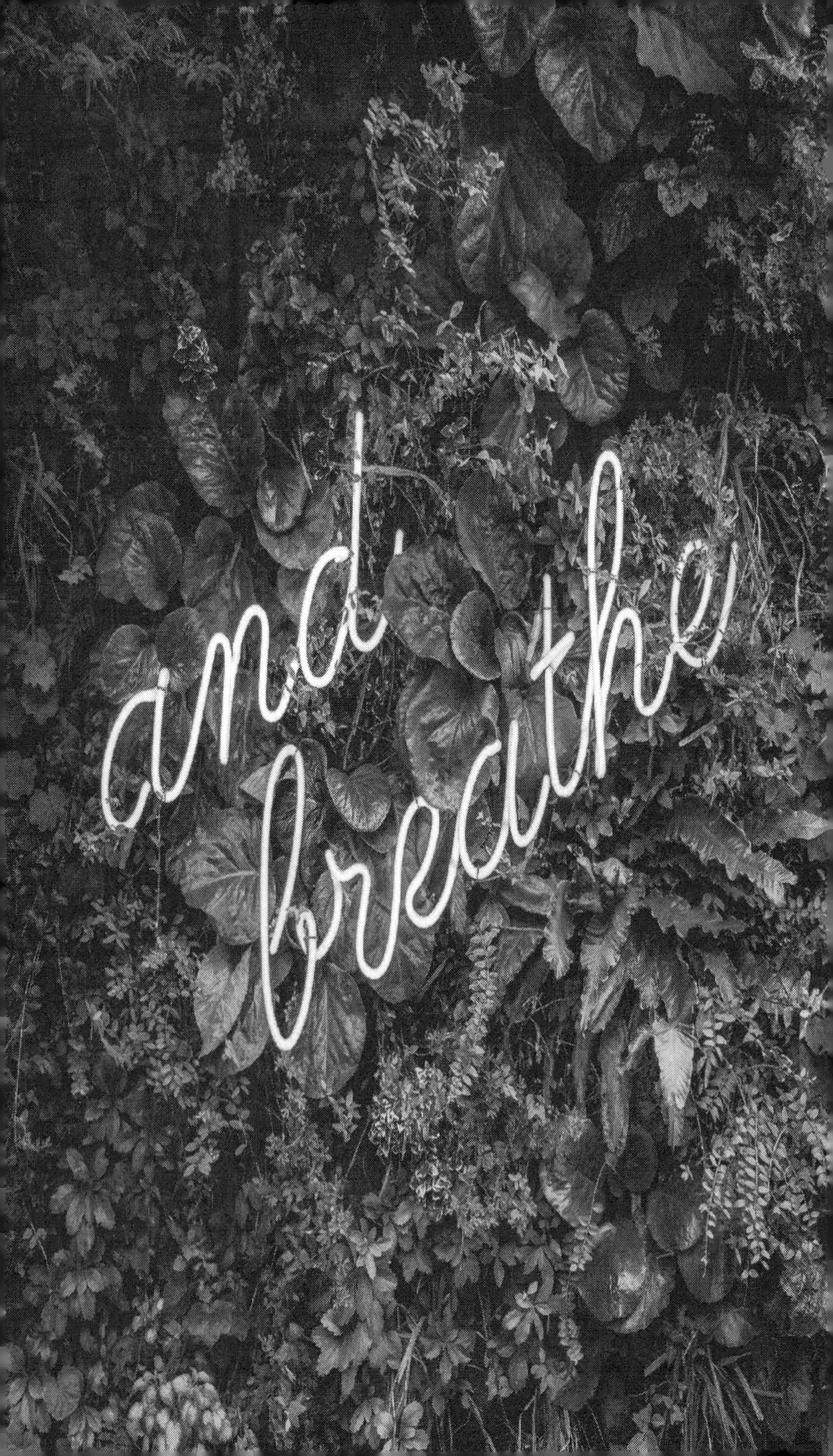
and
breathe

PART 5: CONCLUSION

20. Conclusion

In this book I've sought to take the reader on a tour of the energy transition. We've discussed some of the challenges and opportunities of the sector highlighting some of the opportunities available I the coming years.

We started the story with a view of how energy can be generated and then moved through into transportation and utilisation – finally making some bold calls as to what energy will look like a decade from now.

Energy is an extremely complex topic and I have sought to energise the reader with ideas rather than drown them in detail. Hopefully I have been successful in my quest.

The exciting thing about energy is that there really is no perfect answer and even with the clearest of foresight it remains impossible to predict where the sector will land.

Thank you for taking the time to join me on this journey and I look forward to seeing how these have developed in the 2021 edition.

21. Join the discussion

All comments welcome at:
www.johnarmstrong.co.uk/futureofenergy

Follow me on LinkedIn:
www.linkedin.com/in/johnmichaelarmstrong

Scan with the LinkedIn app:

22. Energy units

Energy is typically measured in watts. Unfortunately, one watt is an incredibly small measure and so to make energy numbers useable engineers use multiples of a thousand watts to talk about energy (Kilo, Mega, Giga and Tera).

To make matters even more confusing we then distinguish between 'instantaneous' energy use and total energy used. Where a total is required, hours are added to the units.

Hopefully the list below helps make this clearer.

Watts
A typical energy saving light bulb uses around 6 watts (6w). Running the light bulb for a day uses 144Wh (which is more commonly described as 0.14 kWh).

Kilowatts (kw): 1,000 Watts - Most commonly used for domestic energy.
A domestic oven uses 3kW. Running the oven for two hours uses 6kwh.

Megawatts (MW): 1,000 Kilowatts - Most commonly used for power generation and distribution.
A gas power station is 450MW. Assuming full output the power station generated 450MWh in one hour.

Gigawatts (GW): 1,000 Megawatts - Used to describe national demand.

The typical instantaneous demand for electricity in the UK is around 40 gigawatts (GW).

Terawatts (TW): 1000 Terawatts - A big number. Used to describe country level consumption.

Annual UK electricity consumption is 310 Terawatt Hours (TWh)

23. Useful resources

The following provide some excellent and regularly updated information:

Bloomberg New Energy Finance: about.bnef.com

Carbon Footprints: www.carbonfootprint.com

Energy Central: www.energycentral.com

Global Carbon Atlas: www.globalcarbonatlas.org

International Energy Agency: www.iea.org

National Grid Future Energy Scenarios:
fes.nationalgrid.com

Shell Energy Scenarios: www.shell.com/energy-and-innovation/the-energy-future/scenarios.html

Sustainable Energy Without the Hot Air:
www.withoutthehotair.co.uk

UK Electricity Grid status:
www.gridwatch.templar.co.uk

World Economic Forum: www.weforum.org

24. Photo credits

All images are courtesy of the wonderful unsplash (www.unsplash.com) or are the author's own.

25. About the author

John Armstrong is an engineer whose career has spanned the extremes of the energy industry – giving him a front seat on the energy rollercoaster. He began his career constructing oil refineries before moving to work across fossil and renewable electricity generation. More recently John has been leading the growth of decentralised energy and district heating in the UK. John is a Fellow is the Institute of Mechanical Engineers, a member of the Energy Institute and holds an MBA in Global Energy from Warwick Business School.

John lives with his wife and two girls in Bath in the United Kingdom.

26. References

[1] www.c2es.org/content/international-emissions/

[2] www.globalcarbonatlas.org/en/CO2-emissions
www.statista.com/chart/16292/per-capita-co2-emissions-of-the-largest-economies/

[3] www.gov.uk/government/news/uk-becomes-first-major-economy-to-pass-net-zero-emissions-law

[4] www.gov.uk/government/news/uk-becomes-first-major-economy-to-pass-net-zero-emissions-law

[5] www.c2es.org/content/international-emissions/

[6] calculator.carbonfootprint.com/

[7] www.huhtamaki.com/en/highlights/responsibility/taking-a-closer-look-at-the-carbon-footprint-of-paper-cups-for-coffee/

[8] gwec.net/global-wind-report-2019/

[9] www.gov.uk/government/collections/energy-trends

[10] www.woodmac.com/press-releases/global-wind-power-capacity-to-grow-by-112-over-next-10-years/

[11] www.gridwatch.templar.co.uk

[12] www.carbonbrief.org/analysis-record-low-uk-offshore-wind-cheaper-than-existing-gas-plants-by-2023

[13] data.gov.uk/dataset/9238d05e-b9fe-4745-8380-f8af8dd149d1/solar-photovoltaics-deployment-statistics

[14] britishrenewables.com/15mwp-solar-park-will-power-the-uks-first-carbon-negative-business-park/

[15] www.tidalenergy.eu/tidal_energy_uk.html

[16] www.edenproject.com/eden-story/behind-the-scenes/eden-geothermal-energy-project

[17] www.weforum.org/agenda/2017/10/fossil-fuels-will-dominate-energy-in-2040/

[18] www.world-nuclear.org/information-library/facts-and-figures/reactor-database.aspx

[19] www.iea.org/fuels-and-technologies/bioenergy

[20] www.electricmountain.co.uk/Dinorwig-Power-Station

[21] www.dominionenergy.com/company/making-energy/renewable-generation/water/bath-county-pumped-storage-station

[22] www.energylivenews.com/2019/12/02/uk-energy-storage-sector-sees-massive-growth/

[23] www.theguardian.com/environment/2019/aug/06/uk-risks-losing-out-europe-home-battery-boom-report-warns

[24] theenergyst.com/the-positives-of-negative-power-pricing/

[25] www.gov.uk/government/groups/heat-in-buildings

[26] www.theccc.org.uk/wp-content/uploads/2018/11/Hydrogen-in-a-low-carbon-economy.pdf

[27] www.bcg.com/en-gb/publications/2019/real-promise-of-hydrogen.aspx

[28] www.economist.com/media/globalexecutive/black_swan_taleb_e.pdf

[29] www.health.ny.gov/environmental/emergency/chemical_terrorism/ammonia_general.htm

[30] www.royalsociety.org/topics-policy/projects/low-carbon-energy-programme/green-ammonia/

[31] www.theguardian.com/environment/2019/dec/25/2020-set-to-be-year-of-the-electric-car-say-industry-analysts

[32] www.mckinsey.com/industries/automotive-and-assembly/our-insights/charging-ahead-electric-vehicle-infrastructure-demand

[33] wardsintelligence.informa.com/WI058630/World-Vehicle-Population-Rose-46-in-2016

[34] blog.aboutamazon.com/sustainability/go-behind-the-scenes-as-amazon-develops-a-new-electric-vehicle

[35] www.eta.co.uk/2017/09/15/electric-hgvs-with-overhead-power-lines-get-go-ahead/

[36] www.business-live.co.uk/economic-development/final-bill-electrifying-great-western-17948591

[37] www.forbes.com/sites/ericrosen/2018/09/08/over-4-billion-passengers-flew-in-2017-setting-new-travel-record/

[38] www.atag.org/facts-figures.html

[39] www.forbes.com/sites/marisagarcia/2018/10/24/iata-raises-20-year-projections-to-8-2-billion-passengers-warns-against-protectionism/#4b6ae712150f

[40] www.eviation.co/alice/

[41] www.cei.washington.edu/education/science-of-solar/battery-technology/

[42] batteryuniversity.com/learn/archive/comparing_battery_power

[43] simpleflying.com/ryanair-25-minute-turnaround/

[44] www.heathrow.com/company/about-heathrow/company-information/facts-and-figures

[45] www.iea.org/reports/tracking-transport-2019/aviation

[46] www.thetimes.co.uk/article/spaces-for-40-000-cars-under-heathrow-s-green-expansion-78t6m2ghq

[47] www.carbonbrief.org/analysis-why-the-uks-co2-emissions-have-fallen-38-since-1990

[48] www.ofgem.gov.uk/ofgem-publications/100628

[49] www.gov.uk/guidance/heat-networks-overview

[50] www.theade.co.uk/case-studies/visionary/islington-councils-bunhill-heat-and-power

[51] assets.publishing.service.gov.uk/government/uploads/system/uploads/attachment_data/file/766109/decarbonising-heating.pdf

[52] www.nationalgrid.com/heating-our-homes

[53] www.theccc.org.uk/wp-content/uploads/2019/07/CCC-2019-Progress-in-reducing-UK-emissions.pdf

[54] energysavingtrust.org.uk/home-energy-efficiency/energy-performance-certificates

[55] www.tado.com/t/en/uk-homes-losing-heat-up-to-three-times-faster-than-european-neighbours/

[56] assets.publishing.service.gov.uk/government/uploads/system/uploads/attachment_data/file/794998/Converting_a_town_to_low_carbon_heating.pdf

[57] www.gov.uk/government/collections/house-building-statistics

[58] carbonintensity.org.uk/

[59] www.nationalgrideso.com/information-about-great-britains-energy-system-and-electricity-system-operator-eso

[60] www.raeng.org.uk/publications/reports/space-weather-full-report

[61] www.ofgem.gov.uk/ofgem-publications/37681/sectoralinvestigations-36.pdf

[62] www.gov.uk/government/collections/energy-and-emissions-projections

[63] www.thetimes.co.uk/article/power-grid-breaks-coal-free-record-with-clear-skies-and-closed-factories-5hzbsfhmq

[64] www.iea.org/reports/renewables-2019/heat

[65] www.resolutionfoundation.org/comment/after-brexit-the-uk-could-cut-vat-on-energy-but-should-it/

[66] www.tado.com/t/en/uk-homes-losing-heat-up-to-three-times-faster-than-european-neighbours/

[67] heatinginstallerawards.co.uk/2018/04/eu-study-show-energy-saving-potential-thermostatic-radiator-valves/

[68] www.tidalenergy.eu/tidal_energy_uk.html

[69] www.gov.uk/guidance/heat-networks-overview

[70] www.thetimes.co.uk/topic/air-pollution